Erica Ferreira
Karla Correia

Population genetics game as a facilitator of the teaching process

Erica Ferreira
Karla Correia

Population genetics game as a facilitator of the teaching process

How to teach population genetics in secondary school in a simple and fun way with a board game

ScienciaScripts

Imprint

Cover image: www.ingimage.com

This book is a translation from the original published under ISBN 978-3-330-76127-8.

Publisher:
Sciencia Scripts
is a trademark of
Dodo Books Indian Ocean Ltd. and OmniScriptum S.R.L publishing group

120 High Road, East Finchley, London, N2 9ED, United Kingdom
Str. Armeneasca 28/1, office 1, Chisinau MD-2012, Republic of Moldova, Europe
Printed at: see last page
ISBN: 978-620-8-36385-7

I dedicate this work to my family.

"It's not machines that

drive and drag the world,

but ideas."

Victor Hugo

Foreword

With a proposal to introduce changes and innovations into the classroom in order to facilitate the teaching-learning process for students, games are an excellent teaching resource. This is an interesting strategy for teaching complex content that is difficult to understand, such as genetics. In particular, the teaching of population genetics, a sub-area of genetics, represents one of the most difficult basic units, as students need to think in terms of populations that change over time. The aim of this study was therefore to design, make and evaluate a didactic board game that would help students understand and learn the content of Population Genetics. [a]This methodology was used with secondary school students from a state school in Anápolis-GO. To assess the effectiveness of the game, the class was divided into two groups, both of which had lectures and only one of which had the game as a learning aid. Data was collected using questionnaires. Both groups answered questions related to the content. However, the first group was asked additional questions about the lecture and the second about the game. The results showed that the students who took part in the game got more questions right. In addition, most of them were satisfied with the game. With this work it was possible to facilitate the teacher's presentation of population genetics content and the students' understanding of it in an efficient and fun way, with their active participation in the construction of knowledge.

Thank you

My sincere thanks ...

... to God for another victory in my life, because without his help, nothing would be possible;

... to my family who, despite all the difficulties, believed in me and did their utmost to help me realise this dream;

... Professor Karla de Castro, for guiding me through this study and guiding its development with great wisdom and patience;

... to the students who took part in this work.

I would like to thank all those who helped me, believed in me and encouraged me to carry out this project.

Summary

Introduction

In recent years, the educational structure has been the subject of discussion in an attempt to introduce changes and innovations into the classroom in order to facilitate the teaching-learning process for students in a more interactive way.

In this context, science or biology teaching should enable students to better understand and participate effectively in the world in which they live.

According to the National Curriculum Parameters (Brazil, 2010), more than just providing information, it is essential that biology teaching also focuses on developing competences that enable students to deal with information, understand it, elaborate or refute it, in short, to understand the world and act autonomously in it, making use of the knowledge acquired from biology and technology.

In this way, it is possible to build solid knowledge in a way that is contextualised with the student's reality.

Delizoicov et al. (2002) report that the function of science teaching in primary and secondary schools is to enable students to appropriate the structure of scientific knowledge and its explanatory and transformative potential, in such a way as to guarantee a comprehensive vision that enables a scientific approach to phenomena and situations both inside and outside the school. The aim is for science and its knowledge to be incorporated as culture.

According to Forneris (1997) apud Casagrande (2006), the concepts of biology are indispensable for the self-knowledge of human beings and their relationship with nature, so that humanity can have a viable future, because responsible citizens need not only to understand the biological world, but also the nature of this knowledge.

Genetics is one of the many subjects covered in biology. According to Burns & Bottino (1991) this content tries to explain the mechanism and basis for both the similarities and differences between individuals and also the immense variation found in all forms of life.

According to Oliveira (2008), in order to revolutionise the way in which teachers have been teaching genetics, it is necessary to provoke curiosity, to instigate students to seek an answer that will allow them to acquire knowledge and apply it to their daily lives. This requires a continuous search for innovative activities that give full scope for questioning and clarifying doubts, always seeking to socialise teaching.

The teaching of genetics to secondary school students is often superficial, as it is a difficult subject for the teacher to explain and difficult for the students to understand, given that it is an area that has undergone many changes, both in conceptual and technological terms.

For this reason, having technologies and interactive games as allies in the teaching-learning process is very important, as they make it easier to understand not only this important content, but also so many other subjects.

Purely expository lectures have a low level of teacher-student interaction, hindering learning and forming passive, non-critical citizens.

For this reason, according to Carboni and Soares (n.d.), one strategy for dealing with this problem would be to use teaching resources, such as games, slides and theatres, which not only make the lesson more relaxed, but also develop the students' creativity and motivate them to learn the content, helping in the teaching-learning process.

According to Klautau-Guimarães et al. (2008), it is necessary to provide teachers with teaching and learning strategies, as well as creating new didactic resources, suited to the space and time available in class, which will enable them to work better and overcome the difficulties associated with teaching and learning Genetics, particularly Population Genetics. For this reason, it is important to create diversified methodologies, such as games, to work on and clarify concepts and provide students with active participation in the construction of knowledge.

Games offer an excellent pretext for students to produce and understand situations and, in addition, Campos et al. (2003) state that didactic games can also fill in many gaps left by the process of transmission-conception of knowledge, favouring the construction by students of their own knowledge.

According to Fialho (n.d.), it is very important that there is a relationship with learning, so that it is characterised by the involvement of both the teacher and the student.

In this way, in their own way, both are being inserted into the teaching-learning process and experiencing the pleasure of appropriating and constructing knowledge.

Teachers must be facilitators of knowledge and use the various existing resources to contribute to their students' educational and social development.

As an example of this, Campos et al. (2003) emphasise that games are an ideal tool for learning because, as well as stimulating the student's interest and developing different levels of personal and social experience, they help them to construct new discoveries and are also a pedagogical tool that allows the teacher to conduct, stimulate and evaluate learning.

According to Rizi (2001) apud Candeias et al. (2005) games are characterised by the following aspects:

- the participant is absorbed intensely and completely, in an atmosphere of great enthusiasm;
- a predominance of spontaneity. Despite the rules to be followed, the participant will be able to count on a variety of alternatives and their performance will depend on their willingness and creativity;
- time is limited - the game starts at a certain moment and continues until it reaches a certain end. In the meantime, various events take place, giving the game a dynamic character;
- replayability - games can be played again at any time;
- the space is limited - every game takes place within a previously delimited area, and
- existence of rules - the rules are created as a way of organising time and space, to facilitate the smooth running of the game.

However, Fialho (n.d.) states that some care must be taken when bringing a game into the classroom, such as:

- **Experimenting with games**

It is of the utmost importance that the teacher tests the game before giving it to the students.

It is necessary to check that the questions involved are correct and that the game pieces are complete, in order to avoid unwelcome surprises during the game. When trying out the game, the teacher can define the number of groups and components that can be formed for its realisation.

- **Quick overview of the content mentioned in each game**

The game is usually introduced to students when they already have prior knowledge of the content involved, so before starting the game, it is important for the teacher to talk about the content that will be covered by the game.

- **Checking the rules**

The rules should be very clear and simple, so that the student understands the game well and feels motivated and interested in the game and wants to win.

- **Proposal of activities related to the contents of the games**

It's important for the teacher to draw up some activities in advance that are related to the content developed in the game, so that this has significant value and really helps in the teaching-learning process. However, care must be taken not to overdo the activities, otherwise the student may lose interest in the game.

A Scoring in games

It's an essential item, as it's the main factor that will motivate the student to play, and it will also stimulate and challenge them while they play.

Scoring encourages competition and, not wanting to lose, students strive to play well, get an excellent score and thus win the game.

Therefore, the game, when well planned, is a very efficient teaching resource in the process of building knowledge, as it promotes reasoning, creativity and student learning.

It also allows teachers to diversify their lessons in order to make them more interesting, attractive and fun for their students without losing their objective, which is learning.

Pereira et al. (2010) developed a playful, low-cost and easily reproducible didactic model that facilitated the teaching of the calculations needed to find the allelic, genotypic and phenotypic frequencies present in a population. This model was applied to a 7th year undergraduate Biological Sciences class.

The work consisted of representing a population in a test tube, in which there were balls of two (2) different colours, representing the alleles.

The pair of balls with the same colours represent homozygous genotypes and those with different colours heterozygous genotypes. A questionnaire was then administered, with questions about the didactic model and the content. The result was satisfactory, as all the students liked the way the "population genetics" content was approached.

Klautau-Guimarães et al. (2008) also approached the subject of "population genetics" in a similar way, using different coloured pins to represent the alleles of a population and thus calculating allelic, genotypic and phenotypic frequencies. In addition, they looked at the effect of each evolutionary factor (migration, mutation, genetic drift and natural selection), but these events were focussed on in isolation. In an informal evaluation, it was pointed out that these activities involved the students more and that they all demonstrated that they had understood the content.

There are few didactic methodologies related to "Population Genetics", which is such complex content within Biology. That's why there's a need to develop more work on this subject that makes it easier for both the teacher to teach the class and the student to understand the subject in a simple and uncomplicated way.

CHAPTER 1

The game as a teaching strategy

Genetics is often taught superficially to secondary school students, as it is a difficult subject for teachers to explain and difficult for students to understand.

For this reason, having technologies and interactive games as allies in the teaching-learning process is essential, since they make it easier to understand not only this importantcontent, but also so many other subjects.

According to KLAUTAU-GUIMARÃES et al. (2008), it is necessary to provide teachers with teaching and learning strategies, as well as creating new teaching resources suited to the space and time available in class, which will enable teachers to work with greater quality and overcome the difficulties related to teaching and learning Genetics.

That's why it's important to create diversified methodologies to clarify concepts and give students active participation in the construction of knowledge. According to CARBONI and SOARES (n.d.), the use of teaching resources not only makes lessons more relaxed, but also develops students' creativity and motivates them to learn the content.

Games are excellent teaching resources that make students understand and produce situations.

According to CAMPOS et al. (2003), educational games can also fill in many of the gaps left by the knowledge transmission-conception process, favouring the students' construction of their own knowledge.

Teachers must be facilitators of knowledge and use the various existing resources to contribute to their students' educational and social development.

As an example of this, Campos et al. (2003) emphasise that games are an important tool in learning, because in addition to stimulating the student's interest, developing different levels of personal and social experience, they help them to build their new discoveries and are also a pedagogical tool that allows the teacher to conduct, stimulate and evaluate learning.

According to FIALHO (s.d), in order to achieve good results using educational games, it is necessary to take certain precautions, such as: trying out the game before applying it, presenting activities related to the content mentioned in the game, clear and simple rules and having scores to motivate the student to play.

Therefore, the game, when well planned, becomes a very efficient teaching resource in the teaching-learning process, as it promotes the student's reasoning, creativity and learning and also allows the teacher to diversify their lessons in order to make them more interesting, attractive and fun without losing their objective, which is learning.

There are few methodologies related to "Population Genetics", which is such a complex subject within Biology.

The aim of this work was therefore to develop and evaluate the effectiveness of a board game as a facilitator, both for the teacher when teaching the lesson and for the student when understanding the subject in a simple and uncomplicated way.

CHAPTER 2

How the game was developed

The research was carried out in a state school located in the municipality of Anápolis, Goiás, with third-year high school students aged between 16 and 18.

After surveying the literature on population genetics to get a better theoretical basis, a board game (figure 1) was made for each group of four students.

The materials used were: Styrofoam, TNT, EVA, a hot glue gun, pins and dice with numbers and figures.

With one Styrofoam plate it was possible to make two sets. Each plate was covered with TNT.

He then cut out the EVA, in different colours, into squares of approximately 5 cm. In these squares, there is a symbol, letter or number, which are also represented on explanatory cards made from EVA (figure 2), which indicate what should be done if the player stops in these places.

These squares represent the path that a population takes throughout their lives, until they reach some place, in this case a cave, where they can rest when they are weak, near the end of their existence.

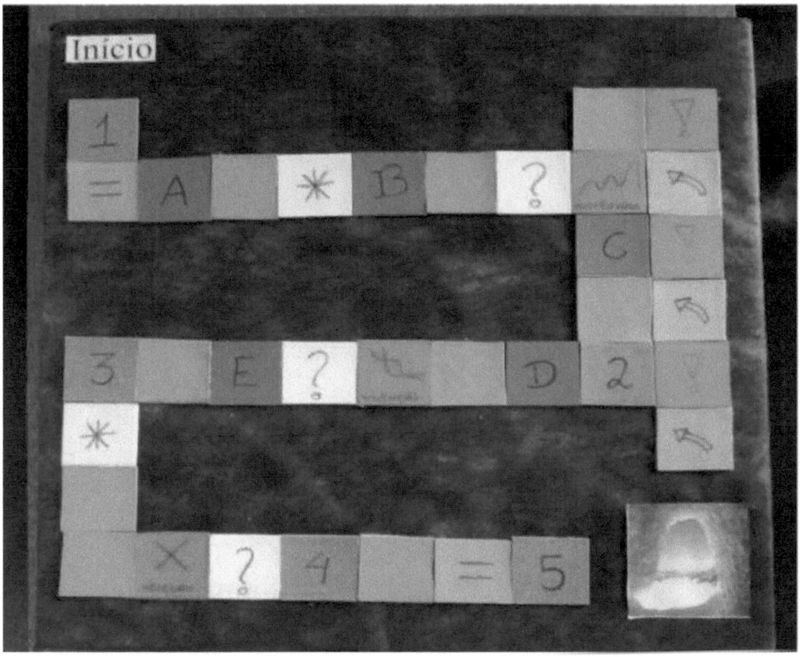

Figure 1: "Population genetics" board game

Figure 2: Game cards

Firstly, the subject was covered in the form of a lecture (3 lessons of 50 minutes each) by the researcher, who was also the class teacher. A total of 30 students took part in this study, 15 of whom only took part in the lecture and 15 in the game. The latter was divided into groups of four (4) students (3 players and 1 assistant to supervise the game). However, due to the number of students, it was necessary for 2 groups to have a single assistant to supervise both games, but this did not jeopardise its application.

Each player received a small Styrofoam plate and randomly chose eight pins of two different colours, light and dark, representing the recessive and dominant gene respectively. The pins represented the alleles and each pair of pins represented the individuals in the population with their genotype. The different alleles should have homozygous genotypes in the same colours and heterozygous genotypes in different colours.

The Styrofoam plate helped to visualise the individuals in the population and to move around during the game (figure 3).

Figure 3: Styrofoam plates with the pins representing the alleles

In order to avoid the risk of the student passing round the board without stopping at some important points, before starting the game, each player must throw the dice with the symbols twice. The symbol on which the die stops are the mandatory stops. The student who reaches the cave first wins the game.

For the "squares" in the game where there is a symbol or figure, there are explanatory cards so that the student knows what to do. The meaning of each symbol can be found in Annex 3.

The rules of the game allow students to better understand speciation and the factors that cause changes in gene frequency such as mutation, selection and migration, and they can also put into practice calculations about allele frequency and whether the population is in equilibrium or evolving.

The game was evaluated using a questionnaire with open, closed and multiple choice questions about the content and the game. The control group, which only took part in the lecture, also answered a questionnaire with the same questions about the content and others in which the students gave suggestions for methodologies to explain some of the content and their opinions about the didactic games.

After collection, the data was tabulated and interpreted using tables and graphs, comparing the two groups in terms of the number of correct answers, errors and omissions.

the efficiency of the game

In general, the didactic game on "Population Genetics" had a satisfactory result, given that the difficulty of assimilating genetics content is a problem faced by many students. They were very receptive and interested, which facilitated the implementation of the methodology. Observation showed that the game provided the students with a dynamic and fun way of learning the content, as well as greater interaction between them. In this sense, it can be said that

> "[...] the game is gaining ground as the ideal tool for learning, insofar as it stimulates the student's interest, develops different levels of personal and social experience, helps them build their new discoveries, develops and enriches their personality, and symbolises a pedagogical instrument that takes the teacher to the position of conductor, stimulator and evaluator of learning (CAMPOS et al., 2003 p.48)."

However, according to BEZERRA et al. (2010) total success cannot be expected from the application of any didactic game that facilitates learning, as each student learns and stores the various contents differently.

The percentage of correct answers is shown in Table 1. Regarding the questions relating to the content, there were more errors and missing answers in the questionnaires answered by the students who only had lectures. This shows that the game helped the students understand the content better.

Table 1: Percentage of correct answers to the questions on "Population genetics".

Question	1	2	3	4	5	6	7
Lectures	66,6 %	60%	86,6 %	26,6 %	53,3 %	93,3 %	46,6 %
Lecture + game	100 %	86,6 %	93,3 %	86,6 %	73,3 %	86,6 %	93,3 %

The majority of students (80 per cent) said that they found the "Genetics of Populations" content more or less easy and 90 per cent said that the subject had been covered previously in an excellent or good way. However, approximately 67 per cent of the students who had only been taught in lectures said that this methodology alone made it possible to learn the content more or less.

Given this data, it is clear that students find it difficult to learn about population genetics, even if the subject has been covered well in lectures.

Of the students who didn't take part in the game, 60% believe that another way of learning this content is through games and 26.6% suggested other methodologies such as: one-to-one lessons, more dynamic lessons and slides. Only 6.6% suggested music and 6.6% suggested theatre.

Of the students who took part in the "Genetics of Populations" board game, 80 per cent said that with the help of the game they were able to assimilate this content better, while the rest (66.6 per cent) thought that it was easier to learn with the help of a game, 33.3 per cent thought maybe and only 6.6 per cent said no.

Thus, it was possible to observe that the majority of students, whether they took part in the game or not, believe that with the help of a didactic game it is possible to better assimilate the content of "Population Genetics".

According to Weller (1995) pedagogical models, such as games, for example, have the aim of "promoting education" and the function of transforming "scientific knowledge into school knowledge", which is one of the reasons why games and other didactic models are difficult to develop, especially in the area of genetics teaching.

In the group of students who only learnt the content through lectures, 93.3% found it interesting that the teacher used other methodologies to explain the subject and 80% thought that a game was a fun way of learning content, as well as judging a didactic game as: interesting, stimulating, motivating, cool and useless, as shown in graph 1.

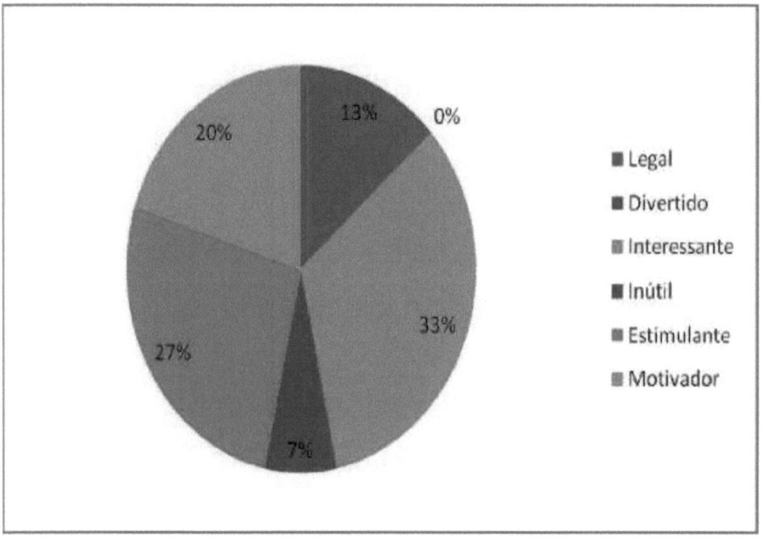

Graph 1: Students' opinions on the possible use of a game to explain difficult content.

With regard to the board game, the students who took part rated it between good and excellent and 60 per cent considered it easy and 86.6 per cent said that the game met their expectations of the content taught. The average score, on a scale of 0 to 10, for the structure of the game was 8.6 and for the way the content was approached was 9.4.

When we asked the students what they thought of the game, the responses were positive, such as:

and "It's a different way of learning as well as being effective and fun"

J "The game makes it easier to understand the subject, we learn by playing",

J "Well-designed, which made us pay attention and take an interest in the subject."

All the students felt that the game helped them learn in a fun and relaxed way. This result corroborates PATRIARCHA-GRACIOLLI (2008) when it says that playfulness in school activities can help in the teaching and learning process, making the student learn in a pleasurable way.

ESQUISSATO et al. (2007) also say that when teachers use didactic models in teaching and learning processes, they provide students with educational moments that enable them to experience new situations at all levels, as well as helping them to understand the content.

FIGUEIREDO (2011) states that teachers who really interact with their students in a fun and relaxed way will get better learning results through play, because when they are emotionally involved in the action, the teaching and learning process becomes easier and more dynamic.

The educational function of the game was clearly observed

during its application with the students, as it favours the construction of knowledge and better assimilation of the content in a relaxed atmosphere.

In addition, RIEDER et al. (2005) state that educational games are playful activities that have specialised pedagogical objectives for the development of young people's reasoning and learning. The game also stimulates student participation, as they feel challenged to play and win.

The use of educational games combined with lectures can generate better results than traditional lectures, as well as providing students with a more dynamic and interactive lesson.

CHAPTER 4

The importance of games in the teaching-learning process

Games are an important methodological tool for teachers to teach and for students to better understand any content. It is a facilitating, interesting and very effective method for the teaching and learning process in the school environment.

With the help of a didactic game, the teaching-learning process becomes simpler and more effective. As well as helping the teacher to explain the content, the game also facilitates assimilation by the student and allows for greater interaction between the students and between them and the teacher.

The playful and educational functions of the game were present throughout its application, and the students were very enthusiastic, as this resource motivated them to learn this content. This shows that the game developed was effective in the teaching-learning process.

The "Genetics of Populations" board game is a simple methodology, easy to make, use and handle by students and teachers in school practice. In addition, it proved to be very effective, as it facilitated the understanding of this content considered "more or less easy" by the students. This game does not require a laboratory or sophisticated equipment, which means it can be used in any school environment.

The creation of this game provides teachers with a new way of teaching "Population Genetics", making this subject simpler and easier to study. This model has proved that there are simpler ways of working on genetics subjects, however difficult they may be, with a good understanding by the students.

With the effectiveness of this "Population Genetics" board game, it is hoped that it has contributed to the appropriation of knowledge and demonstrated the importance of teaching resources, and that teachers will feel motivated to create new teaching strategies that attract the student's attention and interest.

CHAPTER 5

BIBLIOGRAPHICAL REFERENCES

BRAZIL. Secretariat of Secondary and Technological Education. **Parâmetros Curriculares Nacionais - Ensino Médio.** 2010 Available at:< http://portal.mec.gov.br/seb/arquivos/pdf/ciencian.pdf> Accessed on 07 July 2010.

BURNS, G.W.; BOTTINO, P.J. Introduction to Genetics. In: ____________________________________ . **Genetics. 6ª Edition.** Rio de Janeiro: Guanabara Koogan, 1991, p. 1.

BEZERRA, N. P. A. et al. **Elaboration, use and evaluation of didactic games for teaching genetics to secondary school students.** X Conference on Teaching, Research and Extension. UFRPE: Recife, 2010.

CAMPOS, L.M.L.; BORTOLOTO, T.M.; FELÍCIO, A.K.C. **The production of didactic games for teaching Science and Biology: a proposal to favour learning.** [online], 2003. Cadernos dos Núcleos de Ensino, São Paulo, p. 35-48, 2003. Available at:

http://www.unesp.br/prograd/PDFNE2002/aproducaodejogos.pdf Accessed on: 05 July 2010.

CANDEIAS, J.M.G; HIROKI, K.A.N.; CAMPOS, L.M.L. **Use of didactic games to teach microbiology in primary and secondary schools,** 2005. Available at : www.unesp.br/prograd/PDFNE2005/.../autilizacaodojogo.pdf Accessed on 21 November 2010.

CARBONI, P. B.; SOARES, M. A. M. **Molecular genetics in secondary education.** [Available at : http://www.diaadiaeducacao.pr.gov.br/portals/pde/arquivos/1041-4.pdf Accessed on: 21 February 2011.

CASAGRANDE, Grasiela de Luca. **Human genetics in biology textbooks.** 121 f. Monograph (Master's Degree in Scientific and Technological Education). Postgraduate Programme in Scientific and Technological Education. Federal University of Santa Catarina, Florianópolis, 2006. Available at: <http://www.ppgect.ufsc.br/dis/36/dissert.pdf> Accessed on 05

July 2010.

DELIZOICOV, D.; ANGOTTI, J.A.; PERNAMBUCO, M.M.; SILVA, A.F.G. (collaboration). **Teaching Science: foundations and methods.** São Paulo: Cortez, 2002, p. 69.

ESQUISSATO, G. N. M; ARRUDA, G; SOARES, M. A. M. **Didactic model for the Rh factor.** In: Encontro Internacional de Produção Científica Cesumar, 5. Maringá: Centro Universitário de Maringá, 2007.

FIALHO, N. N. **Pedagogical games as teaching tools.** Available at: http://www.pucpr.br/eventos/educere/educere2008/anais/pdf/293114.pdf Accessed on 21 February 2011.

FIGUEIREDO, A. D. L. **New technologies in science teaching.** In: Biological Sciences Degree: module 7, block 1. Goiânia, GO. Funape:Ciar, 2011. p. 25 - 48.

KLAUTAU-GUIMARÃES, M.N. et al. **Dynamics of pins in the teaching of population genetics.** Revista Genética na escola, ano 03, v.02, p. 42-46 ,2008. Disponível

em:
http://www.geneticanaescola.com.br/ano3vol2/10.pdf Accessed on 09 July 2010.

OLIVEIRA, E. B. **Analysis of teachers' and students' perceptions of the Genetics content applied in the 3rd grade[a] of high school in two public schools in the municipality of Apucarana-PR**. Monograph (Degree in Biological Sciences). Faculty of Apucarana - Paraná, 2007. Available at:
http://www.diaadiaeducacao.pr.gov.br/diaadia/diadia/arquivos/File/conteu do/artículos teses/2010/Biologia/monografias/percepcao docente.pdf Accessed on 04 July 2011.

PATRIARCHA-GRACIOLLI, S. R.; ZANON, A. M.; SOUZA, P. R. "Game of predators": a playful proposal to favour learning in science teaching and environmental education. Revista Eletrônica do Mestrado em Educação Ambiental, Rio Grande, v. 20, p.202-216, jan to jul 2008.

PEREIRA, J. S. L. et al. **Elaboration and use of a Didactic Model in the Teaching of Population Genetics.** X Conference on Teaching, Research and Extension. UFRPE: Recife, 2010.

RIEDER, R.; ZANELATTO, E. M.; BRANCHER, J. D. **Observation and Analysis of the Application of Two-dimensional Educational Games in a**

Open Environment. Infocomp. Revista de Ciência da Computação, Lavras, v. 4, p. 63-71, 2005.

WELLER, H. G.1995. **Diagnosing and Altering Three Aristotelian Alternative Conceptions in Dynamics: Microcomputer Simulations of Scientific Models**. *Journal of Research in Science Teaching, 32* ,n° 3, 271290.

CHAPTER 6

ANNEXES

ANNEX 1: Questionnaire for students who only had the lecture

1. What does "Population Genetics" study?

____

____

____

2. What does the expression "gene pool" mean?

____

____

____

3. What is speciation?

( ) It's the formation of a new species.

( ) It is the crossing of different species.

( ) It is the mating of a species.

4. When is a population in equilibrium and when is it evolving?

____

____

____

5. Tick the evolutionary factors that can interfere with the evolution of a population.

( ) mutation

( ) natural selection

( ) population size

( ) migration

( ) female fertility

6. What formula is used to find out if the population is in balance?

( ) $p^2 + 2pq + q^2 = 1$

( ) $p + 2pq + q = 1$

( ) $p2 + pq + q2 = 1$

7. **(VUNESP)** Choose the correct complement for the sentence below. Two populations of the same species, living in different environments and geographically isolated:

a) may form two species if isolation persists.
b) will necessarily have the same gene pool.
c) will not be able to change their gene pools over time because they are isolated.
d) can never form different races
e) will have to be extinguished.

8. How easy do you find the content "Population Genetics"?

( ) Yes ( ) More or less ( ) No

9. In your opinion, how was this content approached during the lessons?

( ) Good ( ) Fair ( ) Poor

10. Was it possible to understand the content clearly through lectures alone?

( ) Yes ( ) More or less ( ) No

11. Apart from lectures, what would be another way for you to learn the content?

( ) Music ( ) Play ( ) Theatre ( ) Other

If you ticked other boxes, write them down:

12. With the help of a game on "Population Genetics", do you think it would be possible to assimilate this content better?

( ) Yes ( ) Maybe ( ) No

13. Do you find it interesting that the teacher explains content using different methodologies?

( ) Yes ( ) Maybe ( ) No

14. How would you feel if your teacher used a game to explain some very difficult content?

( ) Cool	( ) Useless
( ) Fun	( ) Stimulating
( ) Interesting	( ) Motivator

15. What is your opinion of educational games?

( ) It's a fun way to learn content.

( ) It's an unnecessary methodology, because the lecture is enough to learn the content.

( ) With the game it is possible to better understand a content

ANNEX 2: Questionnaire for students who were given a lecture and played the game

1. What does "Population Genetics" study?

____

____

____

2. What does the expression "gene pool" mean?

____

3. What is speciation?

( ) It's the formation of a new species.

( ) It is the crossing of different species.

( ) It is the mating of a species.

4. When is a population in equilibrium and when is it evolving?

5. Tick the evolutionary factors that can interfere with the evolution of a population.

( ) mutation

( ) natural selection

( ) population size

( ) migration

( ) female fertility

6. What formula is used to find out if the population is in balance?

( ) $p^2 + 2pq + q^2 = 1$

( ) $p + 2pq + q = 1$

( ) $p2 + pq + q2 = 1$

7. **(VUNESP)** Choose the correct complement for the sentence below. Two populations of the same species, living in different environments and geographically isolated:

a) may form two species if isolation persists.

b) will necessarily have the same gene pool.
c) will not be able to change their gene pools over time because they are isolated.
d) can never form different races
e) will have to be extinguished

8. How do you rate the content "Population genetics"?

( ) Easy ( ) More or less easy ( ) Difficult

9. In your opinion, how was this content approached during the lessons?

( ) Good ( ) Fair ( ) Bad

10. How do you rate the game?

( ) Good ( ) Fair ( ) Bad

11. With the help of the "Population Genetics" game, was it possible to better assimilate this content?

() Yes ( ) More or less ( ) No

12. Did the game fulfil the expectations of the content covered?

( ) Yes ( ) More or less ( ) No

13. You consider the game easy:

() Yes ( ) More or less ( ) No

14. Have your say on the game.

____

____

____

15. On a scale of 0 to 10, please rate:

a) the structure of the game: _
b) the way in which the content was approached:

ANNEX 3: Meaning of each symbol on the cards

1. You're off to a good start. Move 3 squares!

MO Mountain: The mountain represents a geographical barrier, from here the populations separate. Half must go down the path and half must walk 1 square up. Now the same move applies to both populations.

2. Go back 2 houses!

Hey Mutation: Some animals have reproduced and had offspring with disabilities. Add 2 different coloured pins.

3. Move 3 houses forward!

X The mutation-carrying individuals were negatively selected by the environment and did not survive. Remove them from the game.

Come back to find your gang!!!

! Change the colour of the pins, as speciation has now taken place, i.e. after some time apart, the species have become different from the species in their original group. Go back to the start of the game!

4. Walk 2 houses!!!

5. Go back 2 houses!!!

= This symbol at the start of the game tells you that you must correctly calculate the allele frequency of your population and move forward 2 squares. At the end of the game, all you have to do is say whether your population has remained in equilibrium or is evolving. **NOTE: the frequency of a = 0.4**

? pays gift (each ? is a different colour so that the player knows what his gift is)

ANNEX 4: Consent form

TERM OF FREE AND INFORMED CONSENT FOR THE PARTICIPATION OF THE PERSON AS A RESEARCH SUBJECT

You are being invited to take part in a research project as a volunteer. The data you provide will be kept strictly confidential, maintaining the privacy of the subjects involved. Please note that there will be no payment or financial reward for your participation. If you agree to take part in the study, please sign at the end of this document. If you refuse, you will not be penalised in any way. If you have any questions **about the research**, you can contact the researcher responsible: ________________________________

on the phone: __________ . If you have any questions **about your**

rights as a participant in this research, you can contact ____ no. telephone: **( )** _____________**.**

The subjects below took part in the research of their own free will and were duly informed and clarified by the researcher about the research, the procedures involved, as well as the possible risks and benefits arising from their participation. They were assured that they could withdraw their consent at any time, without any penalty.

If I agree to the terms, I sign below:

MIX
Papier aus verantwortungsvollen Quellen
Paper from responsible sources
FSC® C105338

Printed by Books on Demand GmbH, Norderstedt / Germany